D1456449

THE LIBRARY OF SATELLITES

WEAPONS SATELLITES

PHILIP WOLNY

the rosen publishing group's
rosen
central

To Teresa, for Inspiration

Published in 2003 by The Rosen Publishing Group, Inc.
29 East 21st Street, New York, NY 10010

First Edition

Library of Congress Cataloging-in-Publication Data

Wolny, Philip.
Weapons satellites / Philip Wolny.— 1st ed.
    p. cm. — (The library of satellites)
Summary: Examines the development of weapons satellites which are not yet in use but which, when deployed, can use laser beams to attack large targets, disrupt the weather, or eliminate nuclear missiles in flight.
Includes bibliographical references and index.
ISBN 0-8239-3855-7 (library binding)
1. Space warfare—Juvenile literature. 2. Space weapons—Juvenile literature. 3. Astronautics, Military—United States—Juvenile literature. [1. Space warfare. 2. Space weapons. 3. Astronautics, Military. 4. Artificial satellites.] I. Title. II. Series.
UG1530 .W64 2002
358'.8—dc21
                                                    2002010746

Manufactured in the United States of America

TABLE OF CONTENTS

We normally think of satellites as helping us predict the weather, broadcast television signals and cellular phone transmissions, or gather data on distant stars and planets. But satellites are also up in the sky tracking hostile ships and airplanes that might attack a country, or searching for factories that might be making chemical, biological, or nuclear weapons. At this point, these satellites can only report the information to people on the ground, who will then decide whether to respond with armed planes and ships of their own. Yet satellites are being designed and tested even now that, with the help of high-powered laser beams, will be able to attack large targets, disrupt the weather, eliminate nuclear missiles in flight, or even destroy the satellites of enemy nations. This new breed of satellite is known as weapons satellites.

Many people inside and outside the U.S. armed forces want to create weapons satellites because they feel that a space-based weapons system will be the country's most effective defense against its enemies, especially the new threats posed by unpredictable, rogue nations (such as Iraq) and terrorist groups who seem likely to acquire and use weapons of mass destruction in the near future.

Others think that designing and deploying weapons satellites could backfire on us and that they could cause more problems than they solve. Opponents fear a new arms race, this time occurring in space, which would result in the entire earth being caught in the cross-

Demanding that U.S. president George W. Bush bring an end to U.S. missile defense plans, protesters in South Korea hold a rally in the capital city of Seoul.

hairs of thousands of heavily armed satellites. History will have to judge how we decide. But for good or evil, wars of the future will most likely be waged on land and sea, in the air, and in space.

Welcome to the final frontier of warfare.

CHAPTER ONE

FROM SPACE RACE TO ARMS RACE AND BACK AGAIN

On October 4, 1957, the world was changed forever. On that day, the Soviet satellite *Sputnik 1* became the first human-made object launched into space for the purpose of orbiting Earth. It was made of aluminum and was only 22 inches wide (55.9 cm) and weighed 183 pounds (83 kg). It took *Sputnik 1* about ninety-six minutes to circle Earth completely.

PLAYING CATCH UP

This was a great accomplishment and took scientific achievement to a new level, but it also scared the United States. At the time, the United States and the Union of Soviet Socialist Republics (the USSR, which broke up in the early 1990s into Russia and several other independent states, such as Lithuania and Ukraine) were in the middle of the Cold War, a period spanning roughly forty-five years that began in the aftermath of World War II. This "war" gained its name because the era was marked

A statue of cosmonaut Yuri Gagarin stands next to a replica of *Sputnik* in Moscow's Museum of Cosmonautics. The first man in space, Gagarin left Earth on April 12, 1961. After returning from his one-hour, forty-eight minute flight, Gagarin said that "after looking at the Earth from afar, you realize it is too small for conflict and just big enough for cooperation."

by a strained, but largely peaceful, hostility between two global leaders. Although the two nuclear super-powers never squared off against each other directly in a "hot" war, they did engage in an extremely expensive arms race and got involved in various proxy wars in which they funded or fought alongside the Communist and anti-Communist forces of other countries. Many people thought that the security of the United States and its allies would be put into grave danger if the Soviets could beat the United States into space. The

Explorer 1 entered into an orbit around Earth on January 31, 1958, making it the first U.S. satellite to reach outer space successfully. Its instruments detected large bands of radiation around Earth, the first major discovery of the space age.

United States had to launch its own satellite, and fast. The space race had begun.

In the beginning, attempts by the United States to catch up to the Soviets resulted in embarrassing failures. The rocket launches that would carry the first U.S. satellites into space all ended in disaster. They either exploded right on the launch pad or burst into a million pieces a few miles above Earth. On January 31, 1958, however, *Explorer 1*, a scientific satellite that was designed to gather data on the radiation belts that surround Earth, was sent successfully into the heavens. The United States had finally gained a foothold in space.

SPY SATELLITES

The stage was set for a race not only to gain control of outer space, but also to use dominance in space to gain military advantage on Earth. During this time, each side also attempted to outpace the other in building and stockpiling nuclear warheads, each year constructing more advanced and destructive ones. Missiles would be used to deliver these warheads to their targets. To help monitor each other's nuclear stockpiles, the United States and the Soviet Union both developed spy satellites that would allow them to determine the number, location, and movements of the enemy's warheads. Each nation needed to keep track of the other's conventional weapons, too: armored divisions of tanks, warplanes, aircraft carriers and battleships, even troop movements. The new spy satellites made such monitoring possible. They could not, however, do anything other than report to Earth about what they saw. They were not capable of destroying the arsenals they monitored.

Some of the first spy satellites were fairly crude. They took only conventional pictures and could not see through clouds or use radar. Unlike today's satellites, they could not detect heat, either. To one of these early satellites, an advanced fighter plane looked no different from a harmless cargo plane.

One of the first advances that helped satellites "see" better was the invention of multispectral scanning. This means that the satellite could now take several pictures of the same object, each photograph taken in a different wavelength. This new method allowed spy satellites to tell the difference between titanium, aluminum, or wood, helping to differentiate between different kinds of planes. In addition, new infrared photographic equipment was also created. This meant that satellites could identify the heat of missile launches and booster rockets. The first of these satellites, *Midas 7*, was launched in 1963 and successfully detected a missile launch during a test. The United States could now monitor the Soviet Union's missile-testing programs and receive early warning of any actual missile launches. From now on, it would be difficult for either nation to keep any nuclear secrets.

"STAR WARS"

In addition to creating satellites that could spy on each other, the United States and the Soviet Union were also trying to develop more active satellites, ones that could shoot down intercontinental ballistic missiles (ICBMs). ICBMs can carry nuclear warheads thousands of miles and threaten cities anywhere on the globe. During this time, many people in both countries were concerned that developing antiballistic missiles (or ABMs, the missiles that would destroy incoming ICBMs) would not make the world any safer from nuclear attack, but would instead simply spark another escalation of the arms race. These concerns resulted in the ABM Treaty of

1972, which severely limited the number of these missiles each country could have in its arsenal.

So this early conception of a space-based missile defense system was shelved for the time being. More than ten years later, however, on March 23, 1983, President Ronald Reagan delivered his famous speech outlining a new policy that he believed the federal government should implement: the Strategic Defense Initiative (SDI).

In 1988, three years before the fall of the Soviet Union, reporters were afforded a rare glimpse of a Soviet ICBM as it rested in its underground silo. Luckily enough for both sides, the Cold War ended without any missiles being fired.

SDI was nicknamed Star Wars after the blockbuster movie series created by George Lucas that had captured the American imagination in the 1970s and 1980s. SDI was imagined as a space shield that would utilize satellite technologies, X-ray lasers, and other weapons to intercept any and all nuclear missiles launched against the United States.

Sitting in the Oval Office, President Ronald Reagan speaks to the nation about the controversial Strategic Defense Initiative, which drew fire from its opponents for being costly and possibly ineffective.

SDI did not enjoy universal enthusiasm, however. Critics of the initiative thought that building such defenses would take many years to design, develop, and deploy, and would never be completely foolproof even after all that effort. Many thought it would not be worth the enormous costs that would be incurred, totaling hundreds of billions of dollars. Opponents, many of them in the U.S. Congress (who would vote on whether or not to approve the initiative), also thought that the technology was not yet available to create such a space shield. The Reagan administration and Congress fought over the issue for years.

Tied up in Congress throughout the administrations of Reagan and George H. W. Bush, many of the individual projects that together constituted the SDI were eventually given up on and lost their federal funding.

Dreams of SDI took a backseat to other concerns—such as welfare reform, education spending, and environmental protection—under President Bill Clinton.

AN OLD IDEA CONFRONTS A NEW WORLD

With the election in 2000 of George H. W. Bush's son, George W. Bush, to the presidency, however, the idea of a space-based defense of U.S. interests was resurrected under a new name—National Missile Defense (NMD). Even though federal funding for research and development had been severely cut over the years, some aerospace companies continued their studies in this field. As a result, over the long years in which SDI was debated and ultimately shelved, the technologies needed to create satellites that could launch ABMs kept getting smaller, smarter, and cheaper, making the idea more feasible than it was twenty years earlier when Reagan first proposed it.

The nature and source of the threats that satellite-based missiles are supposed to defend us against have also changed, however. With the fall of Communism in the former Soviet Union in the early 1990s and the growing friendship between the United States and Russia, supporters of NMD look to more current threats in the world to justify the program. The terrorist attacks against the United States on September 11, 2001, for example,

BREAKING WITH THE PAST

In December 2002, President Bush announced that America would be backing out of the ABM Treaty with Russia. According to a White House press release, he said, "The ABM Treaty hinders our government's ability to develop ways to protect our people from future terrorist or rogue-state missile attacks." Many Americans and many of our allies were troubled by this move. The Bush administration had signaled its determination to deploy space-based defensive weapons, even if it meant breaking treaties to do so.

came from a very different kind of enemy than a Cold War superpower—much smaller, but far less detectable and predictable. Some people believe that a terrorist group like Al Qaeda is no less a danger than the old Soviet Union was, even though it has far fewer weapons and no country that it calls home. It only takes one weapon of mass destruction in the hands of people who feel they have nothing to lose to kill thousands, even millions of people. Some people believe that a terrorist organization could gain control of a missile silo some-where or construct a crude nuclear device and hold the world hostage to its demands.

Free-ranging terrorists are not the only concern, however. More and more countries throughout the world are gaining the knowledge and technical abilities to create nuclear weapons, and not all of these countries

Although pioneered by the United States, nuclear weapons technology has now spread across the world. In Lahore, Pakistan, citizens hold their country's flag aloft while celebrating the successful completion of five underground nuclear tests. Pakistan is stockpiling nuclear weapons in the event of war with India, its regional rival with whom it shares a disputed border.

are friendly or stable. North Korea, one of the last remaining Communist countries and long considered a volatile and unpredictable enemy of the West, is thought to be working on ICBMs that can reach Alaska. India and Pakistan recently began producing and stockpiling nuclear weapons, even as they engage in conventional warfare (warfare waged with non-nuclear weapons) over Kashmir, the Indian state that is claimed by both nations.

The stockpile of nuclear weapons spread throughout the countries that once made up the Soviet Union is still very large. Although many of the former Soviet Republics—most importantly Russia—and the United States are now cautious allies, the Soviet-era nuclear arsenal is thought by many to be vulnerable to terrorists and revolutionaries who could forcibly gain control of it. China, too, is thought to be developing a nuclear weapons program. The United States and other Western nations are trying to establish normal ties with China, but relationships remain tense and strained. It could take only a small incident to have the largest countries suddenly threatening each other with nuclear weapons.

According to its supporters, the NMD is our best protection against these new and disturbing threats. A new space race seems to be on, as the dream of a satellite-based national missile defense is forging ahead over the objections of many critics.

CHAPTER TWO

THE HIGH GROUND AND ITS PERILS

Before World War I, when the first airplanes were mobilized for dogfights over Europe, war was fought entirely on land and sea. Ground war required using a battlefield's topography (natural features) to its best advantage. For example, an army that gained the high ground, such as the top of a hill, had a strong advantage over its enemy below. The soldiers at the top were better able to hit their more exposed targets below with whatever the current weapon of choice was—stones, arrows, boiling oil, bullets, cannonballs, or mortar shells.

But when it came to military intelligence—learning where enemies were, what they intended to do, how well they were armed, and how many they were—information gathering was extremely basic. Spies were sent to sneak behind enemy lines, observe the troops, and report back. Without this information, generals could never be sure if the enemy was hiding in the woods, for example, laying an ambush. Without intelligence gathering, what was seen in plain view was all that could be known.

The Sopwith Camel shot down more enemy aircraft than any other Allied plane during WWI. Arthur Roy Brown, a pilot with the Royal Canadian Air Force, was flying a Sopwith Camel when he shot down German ace Manfred von Richthofen—better known as the Red Baron.

CLIMBING HIGHER AND HIGHER

The airplane changed how intelligence was gathered and wars were conducted. It was used by the military first as a tool for reconnaissance (the investigation of enemy territory) and then as a deliverer of weapons (such as bombs and missiles) that could both attack enemy sites and provide support for ground troops. From World War II to the Vietnam War and on through

Operation Enduring Freedom (the allied effort to root out Al Qaeda terrorists in Afghanistan following September 11, 2001), air superiority became a crucial element in waging war successfully. The nation that controlled the skies could offer its ground forces greater protection and support, and wreak greater devastation on its enemies. As a result, territorial gains often followed air mastery.

The new era of satellite technology that was ushered in with *Sputnik* took things even further. The quest for air superiority began to be expanded ever outward and upward, into space itself. Though unable to fire missiles or drop bombs like fighter jets and bombers, spy satellites could provide highly detailed reconnaissance and very exact information about the location of targets, allowing for extremely precise bombing strikes.

Space is the new high ground. From an orbiting position above Earth, a nation can direct its jets, battleships, missiles, and ground troops with precision and efficiency, all the while keeping careful track of enemy movements. The only problem is that, unlike on land, more than one force can occupy the high ground at the same time. This negates any advantage and can lead to an endless series of attempts to regain the competitive edge. What was meant to be a shield can quickly turn into a space-based arms race, a new target for terrorists, or even worse, an all-out war among the stars.

SPACE-BASED TERRORISM

These days, governments, private companies, and ordinary citizens are dependent on satellites for communications, weather forecasting, military reconnaissance, and scientific research. In many cases, we cannot make a cell phone call, watch television, get on the Internet, e-mail a friend, or send a fax without the help of satellites. The problem is, the more we come to depend on satellites, the more vulnerable we will be if they are threatened. If another country were able to wipe out a portion of the United States's civilian and military satellites, Americans could suddenly find themselves living in a new dark age.

Such a scenario is not far-fetched. Satellites themselves are becoming cheaper and easier to launch. Governments, private companies, and even terrorist groups sponsored by a wealthier nation could put a satellite into space for a few million dollars. Before long, smaller nations and terrorist groups may no longer scramble to build or buy intercontinental ballistic missiles, but instead may try to deploy satellite-mounted lasers that can sink battleships and wreak havoc and destruction on cities.

Satellites themselves—even defensive ones—are very vulnerable to attack. A nation's communications, intelligence-gathering, and defense abilities could be severely compromised by the crudest of methods. Anti-satellite weapons (ASATs) do not have to be incredibly

SALYUT 3

There are no space-based weapons currently in orbit, but that has not always been the case. According to a 1998 article in *Spaceflight*, a magazine published by the British Interplanetary Society, the first example of an armed satellite was the Soviet Union's *Salyut 3*. Launched in 1974, it was the first manned military outpost in space. With a two-man crew, it had a modified 23mm rapid-fire cannon strapped to the outside. This was put there for "defense against U.S. space-based inspectors/interceptors," according to Soviet scientists. A periscope attached to a visor on the control panel let the crew aim the gun. The gun was limited in its abilities, however. First, the entire satellite had to turn around in the direction the astronauts wanted to fire the gun. Not only that, but the force of the gun's recoil could knock the satellite out of orbit. So the *Salyut* was equipped with thrusters that would work only when the gun was fired, helping it to maintain the proper orbit.

expensive or advanced. As James Oberg wrote in *New Scientist* in June 2001, the concept of a "poor man's ASAT" is something dozens of nations have within their power: A small missile can be launched and can "deposit a cloud of sand, ball bearings, and other hard objects in the path of an oncoming satellite. The target's own velocity provides the impact energy." A small piece of debris seems to pose no danger floating freely in space. Put it in the way of a satellite flying through orbit at 17,000 miles (27,359 km) per hour, however, and you have a recipe for disaster!

Satellites equipped with antiballistic space-based lasers are also not foolproof. To overcome a missile defense system, an enemy nation can simply launch a saturation attack, firing off more ballistic missiles than the satellites can possibly take down. Another simple way to foil a space-based missile defense system is to launch nonlethal decoy missiles along with ones bearing nuclear warheads. The satellites would have no way of differentiating them and would have to try to destroy all of them, leading to possibly deadly wasted effort. The nuclear missiles' chances of successfully reaching their target would be improved greatly if the satellites were busy shooting down dummy missiles.

THE DANGERS OF SPACE DEBRIS

One of the greatest perils arising out of arming the heavens is a problem we do not often associate with the clean vastness of space—litter. Since the space age first began, we have been sending up satellites, rockets, space shuttles, space stations, and various other craft into space. Currently, there are more than 8,000 detectable objects (larger than four inches or ten centimeters) in orbit around Earth. Probably six or seven hundred of these are working spacecraft. The rest of these are, more or less, space junk: satellites that have been retired or shut down, have burned out or malfunctioned, or have simply been

When a spacecraft is traveling tens of thousands of miles an hour to maintain an orbit around Earth, a collision with even the smallest particles and debris can be dangerous. This small hole in one of the space shuttle *Endeavour*'s side hatch windows was caused by a very tiny micrometeoroid.

abandoned. Other objects include discarded booster rockets from rocket launches, tools lost by space crews during space walks, even bits of paint.

And these are just the trackable objects. It is estimated that there are up to 150,000 smaller objects (less than ten cm) floating in orbit. Pieces of space junk can cause catastrophic damage to a satellite speeding along at 17,000 mph (27,359 km/hr), and new satellite designs have to take this dangerous debris into account. After the Soviet military tested ASATs in the late 1960s, a dozen clouds of metal shards were left floating in orbit. Now all spacecraft have to be armored against these.

If war in space becomes a reality, then the potential threat from space junk will get many times worse. Nations would be destroying each other's orbiting satellites, adding thousands of new pieces of debris with each "kill." Civilian and military satellites and space stations would likely be severely damaged by collisions with debris, endangering communication, intelligence, meteorological, and data collection services on Earth. Even neutral countries far below would have to worry about larger pieces of debris raining down on them from above. A large explosion could hurl debris out of orbit and back into Earth's atmosphere, where it would begin a fiery descent to the world below. Even a killer satellite attacking an enemy satellite would have to be programmed to take evasive action in order to avoid the resulting explosion and rain of shrapnel.

Worse even than the metallic debris of an exploding satellite, however, is something else that might be released following an attack—radioactive fuel and materials. Weapons satellites are going to be bigger and require more energy than regular, "passive" satellites will (which are powered by the Sun using solar panels). Nuclear power is the most likely source of energy for these weapons satellites. Would you feel safe knowing that radioactive materials could rain down on you after a fierce orbital battle?

CHAPTER THREE

INSIDE WEAPONS SATELLITES

If weapons satellites ever move beyond the drawing board, how would they work? In most respects, they would not differ very much from other satellites that are launched for communications, earth imaging, weather, or reconnaissance purposes. All satellites have certain basic functions in common. The primary difference is that weapons satellites would be "active," not "passive" machines. Rather than simply observing things going on below and sending data back down to Earth, like most satellites, weapons satellites would be reacting to what they see and making life or death decisions in the defense of nations below.

A weapons satellite's specific job would depend on its role within the missile defense network. Some would be equipped with lasers or other weapons and would be responsible for destroying enemy missiles launched against a nation's territory. Some would simply be orbiting mirrors used to direct and strengthen the laser beams fired by other satellites in the network. Others would serve as traditional reconnaissance satellites, helping to detect the presence of missiles or other weapons deployed anywhere around the world. Each type of satellite within the defense network would build upon a variety of technologies. Some of these technologies have been in development for years and are only just now ready to be used in real-life situations.

SATELLITE BASICS

Although satellites differ in many ways, they all have certain aspects in common. All of them have a bus—the metal or composite frame that surrounds all the other parts. This is a sort of shield or protective armor that guards a satellite's sensitive and delicate equipment from the perils of space, such as collisions with space debris, a buildup of electric charge, extreme temperatures, and radiation.

A source of power is necessary to keep a satellite's instruments working. Many use batteries that are recharged with solar energy collected by solar cells attached to the satellite. Newer satellite designs are using fuel cells, which are like rechargeable batteries that use hydrogen and oxygen instead of electricity to repower themselves. Nuclear power, which has been used to power deep-space probes (crafts that venture to the far reaches of our solar system to gather information on its various planets and moons and to peer beyond into the surrounding galaxy), may become a common energy source for satellites in the future. Weapons satellites, in particular, might require it. Their large size and their energy needs would rule out powering them with conventional batteries or solar energy.

Almost all satellites are powered by solar cells, large panels that convert the Sun's light into electrical energy. On a bright, sunny day on Earth, the Sun provides about 1,000 watts of energy per square yard of the planet's surface.

In addition, all satellites require an onboard computer that controls the proper functioning of the satellite and its different systems. Plugged into the onboard computer is the radio system that allows ground crews to check on the status of the satellite and its operations. It also lets them relay instructions to the satellite's computer, directing it to perform certain tasks.

Finally, all satellites have an Attitude Control System (ACS). The attitude refers to a satellite's position or angle in space. To be effective, satellites have to be kept pointed in the proper direction at all times (which is determined by

The United States hopes to have satellites bearing optical lasers in space by 2012. These costly, experimental satellites will be used to destroy enemy missiles before the missiles have a chance to reach their target.

lasers, or SBL. A system of satellites orbiting Earth will have the job of either shooting down or destroying nuclear weapons fired against the United States or other nations. One plan calls for satellites affixed with mirrors (and separate from the SBL satellites) to help guide the lasers by reflecting and directing them straight to their targets. In another design scenario, the mirrors and lasers would be part of the same satellite.

A space-based missile defense network would become a crucial component of the early warning

LASERS ON THE GROUND

The idea of using spaced-based lasers to defend against enemy missile attacks has been discussed and explored for years. Initially, however, laser defenses were designed for use on the ground. One of the first experiments in which lasers were used as defensive weapons took place in 1973. It was then that the Mid-Infrared Advanced Chemical Laser (MIRACL), a deuterium-fluorine laser built by the U.S. Air Force and aerospace contractor Thompson Ramo Woolridge (TRW), was successfully tested against missiles and remote-control drone aircraft.

defense system of the United States and its allies. If it works as designed, it would be able to shoot down dozens, even hundreds, of enemy missiles soon after they have been launched, during the boost phase. It is at this time that a missile is most vulnerable because it is struggling to escape gravity and Earth's thick, heavy atmosphere. In other words, it is under a great amount of stress. The missile is also very easy to detect and track at this time, because of the great amounts of heat generated during launch and the initial boost into space.

The type of laser most likely to be used in any future SBL program would be produced by the mixing of deuterium, nitrogen trifluoride, and helium to produce fluorine. The fluorine would then be burned with hydrogen in an optical resonator—a special chamber lined with mirrors (the mirrors' reflective surfaces intensify laser radiation). The result is a mass of

The LAMP mirror is the largest mirror ever built for use in outer space. Because it will be redirecting lasers over very long distances to very specific targets, it is important that the mirror's surface, measuring 13 feet (4 meters) in diameter, is completely flawless.

"excited" hydrogen fluoride molecules. When these return to their natural state, they release photons. Another optical resonator concentrates and amplifies (expands) these photons and channels them into a laser beam.

MIRRORS

Generating the laser is not enough, however. A laser beam gets weaker the farther it travels. Beyond a certain distance, it will no longer be able to inflict serious damage to its target. To remedy this, each of the SBL satellites needs an optical assembly. This piece of equipment is the equivalent of using a magnifying glass to focus rays of sunshine on dry twigs or leaves, causing them to catch fire. Using special mirrors, the optical assembly will focus and intensify the laser so that it remains powerful even after traveling vast distances.

This focused, directed, and intensified laser beam can then be bounced off an orbiting space mirror and toward an enemy missile, spacecraft, or other satellite. The space mirror provides greater accuracy for the laser. Since Earth is round, and light bends as it moves through space, a satellite would not be able to shoot directly at its target because the laser beam would curve as it traveled to Earth. Instead, it would have to fire its laser at a mirror near the horizon, which would then bounce it directly at the target.

During the 1980s, there was a push in military and civilian labs to create high-grade mirrors to be used for SDI, the original space-based missile defense program. One of the results was the Large Advanced Mirror Program (LAMP), which was finished by 1989. The mirror that was produced is thirteen feet (four meters) wide and segmented, meaning that different parts of it move in different directions. Mechanisms called actuators move each section of the mirror to where it needs to be so that lasers reflected off it go in the proper direction and gain the right amount of power.

ADAPTIVE OPTICS

Have you ever looked up at the sky and wondered why the stars twinkle? It is because of the atmosphere. Gases in the atmosphere are always swirling in motion, and Earth itself spins, too. All this movement affects the way light travels from the stars to us, bending and distorting it so that stars seem to twinkle and the summer sunlight seems to blaze and wave. In the same way, the atmosphere can distort the straight path of a laser traveling through it, changing its direction and weakening its force.

To make up for this, scientists have developed deformable mirrors that can change shape. The system works by first sending a low-powered laser toward Earth. When the laser's radiation is scattered by the atmosphere, it is reflected back to the satellite. The satellite's sensors record exactly how the radiation is scattered. This tells the system how much distortion of the laser there was during its journey to Earth. Based on this information, the deformable mirror bends in a certain way and "preshapes" a laser beam that will make it through the atmosphere in a straight and true line.

ACQUIRING, TRACKING, POINTING

One of the most important features of an SBL satellite will be its ability to fire lasers with accuracy and destructive force. Sensors in the satellite and on the ground—instruments that will locate the precise location of targets, track their movements, and aim the laser—need to be accurate and quick. The Acquisition, Tracking, and Pointing (ATP) system will get the laser where it needs to go. Acquisition means locating the enemy missile once it is fired. Tracking is the process of monitoring its movements and path once launched. Pointing means directing the laser to the enemy missile and keeping the laser there long enough to destroy the target. A fourth part of the system that is important is fire control. This refers to how fast the laser can move to a new target after it has destroyed the previous one.

The United States completed two experiments important to ATP by 1990. The Relay Mirror Experiment (RME) was one of these. In RME, a ground-based, low-power laser was fired at a mirror satellite in low-earth orbit and bounced back to a ground target in another location. This successful test showed that lasers could be accurately directed by space mirrors against a target.

The second important advancement for ATP in the late 1980s was the Rapid Retargeting/Precision Pointing simulator—R2P2, for short. This simulator was actually built on the ground but was made to replicate closely the conditions in space. During tests, the mechanisms for stabilizing the laser beams and maintaining their accuracy passed with flying colors, as did the rapid response of the fire control system.

GETTING SBLS OFF THE GROUND

We have seen how various technologies are being investigated and developed with the goal of defending nations—from high up in space—against enemy attacks. But what is being done right now to get these ideas off the ground and into orbit?

The U.S. Air Force recently teamed up with three aerospace giants—Boeing, Lockheed Martin, and Thompson Ramo Woolridge—to try to set up the first extensive space war experiment, called the Space-Based Laser Integrated Flight Experiment (SBL-IFX). Scheduled to start testing in 2012, the object of the experiment is to test successfully the satellites and weapons that would be used to deter a small, isolated nuclear attack (involving no more than a few missiles) of the sort a terrorist group or unstable developing country might launch. Once in orbit, it's likely that the experiment will test one or all the chemical laser weapons mentioned earlier in the chapter. Test balloons and dummy rockets with sensors will be used as targets. The SBL satellite will have a mirror that unfurls to about 15 feet (4.57 meters).

The costs of researching, building, and testing the SBL system is expected to run well into the billions of dollars. The lowest estimate for an SBL system that features twenty orbiting defense satellites ranges between $17 billion and $29 billion. The higher estimates run as high as $81 billion. Those who oppose the system think that even this figure is unrealistically low.

Given that many people feel these satellites will make us no safer from enemy threats, but may indeed clear the way for new and more horrific kinds of warfare and terrorism, critics of SBL wonder if we would be paying a fortune and receiving nothing but grief in return.

CHAPTER FOUR

THE FUTURE

Although lasers are the current weapons of choice for planned space-based missile defense systems, they are not the only ones being considered. It is conceivable that, within a generation or two, space could be bristling with all sorts of destructive satellite-based weapons, from the sophisticated—such as particle beams and microwaves—to the relatively crude dropping of "air spikes" and manipulation of the Sun's rays to disorient enemies and burn their facilities. There is no shortage of destructive tools that can be placed in space, perhaps the only limits being those of the human imagination. It does indeed seem possible that the wars and violence we are too familiar with on Earth could soon be a common feature of space as well.

PARTICLE BEAMS

Using lasers to shoot down enemy missiles and attack targets on the ground is the most likely route the U.S. Department of Defense will take in developing a weapons satellite. Yet there is an even more destructive alternative: the particle beam. Like a laser, the particle beam can travel at close to the speed of light, yet it contains far more energy. As a result, it can cause greater damage to its target.

To generate a particle beam, hydrogen or deuterium gas is exposed to a great electrical charge. Negatively charged ions are then accelerated through a vacuum tunnel (a place without air or matter in it). The negative ions are stripped of their electrons at the end of the vacuum tunnel, increasing their energy, and are shot out of the weapon at nearly the speed of light. These "bolts" of energy are then directed at their target.

When the beam hits its mark, the results are devastating. The particles fired by the weapon actually enter the materials of the target. The incredible amounts of energy the beam contains are passed onto the atoms of the target itself. The result is a rapid and extremely intense heating of the target that causes it to disintegrate, or break apart. Sometimes the energy of the particle beam is so great that its target actually explodes.

Firing particle beams from a space-based weapons platform is an idea for the distant future, however. Hundreds of millions of volts are required to generate the beam's intense power. Today's orbiting satellites cannot produce anywhere near that much energy. In addition, if developed today, the system would weigh a few hundred

Holding a piece of aluminum with a hole burned through it by a particle beam, Lieutenant Colonel Warren Higgens of the U.S. Army speaks at a press conference on the launch of the Beam Experiment Aboard Rocket (BEAR) on July 17, 1989. Because of the weight of the weapon, however, it is still impractical to put a particle beam into space.

tons, making launches a very expensive proposition. Also, the particle beam, when it is fired, needs to hit a target that is within its line of sight. The beam could not be redirected by mirrors, as lasers can be. Like those of lasers, the paths of particle beams are affected by Earth's atmosphere. Unlike lasers, particle beams destroy anything they are aimed at, including mirrors. Unable to rely on mirrors to redirect their course, engineers still have to find an accurate way to aim and fire a space-based particle beam at its target in the lower atmosphere or on Earth.

MIRRORS

Directing and intensifying lasers are not the only roles imagined for mirrors in space. Plans for the solar energy optical weapon (SEOW) include a group of space-based mirrors that reflect and intensify the Sun's rays, which can then be directed onto ground, air, or space targets.

A slightly different version of the same idea calls for giant mirrors as big as 6 miles (10 km) wide to perform the same job. War planners think mirrors between 10 and 100 meters (32.8 to 328 feet) wide could be used to light up nighttime battlefields, like a giant overhead light burning in space. The larger the mirror, the greater the area on earth that could be illuminated. Other imagined uses include using blinding, reflected light to disorient enemy troops. Some scientists even think that we can affect battlefield weather with redirected sunlight. Mirrors could also be used to melt or burn enemy facilities and satellite components.

MICROWAVES

Another way of turning up the heat, literally, on enemy forces is through the use of microwaves. One example is the space-based, high-power microwave weapon (HPM). Positioned in low-earth orbit (about 500 miles, or 804 km, up), an HPM satellite could direct microwave energy at targets in space up to half a mile away. Microwaves lack the long range of lasers, but they are no less fearsome. Instead of blowing up a passing satellite, a microwave could melt its electronics, making it a floating piece of space junk. Microwaves directed from space would also be useful in disrupting electronic systems on the ground, creating havoc with an enemy nation's communications and computer systems.

MINIATURIZATION

The cost of launching hardware into space is very high, mainly because of the amount of fuel needed to boost a heavy load into orbit. As Alok Das, head of innovative concepts at the Pentagon's Space Vehicle Directorate told the *New York Times*, "It costs a bar of gold to launch a can of coke." Aerospace engineers are working hard to fix this problem. Newer lasers and other weapons are being designed to be lightweight and more powerful. The real cost savings, however, may come from thinking small. Some new designs feature weapons satellites that are smaller than ever imagined before.

One of these is the microsatellite. Unlike current models, the microsatellite will be about the size of a suitcase and weigh close to 200 pounds (90.7 kg). Das, who is heading the microsatellites initiative, says these small satellites will be launched in large groups, which will save money spent on booster rockets and fuel. Each will be able to perform different tasks but can swap jobs when necessary. These smaller satellites could perform surveillance or be used as maintenance craft, refueling or upgrading other, larger satellites. Planners think these microsatellites could send live video feeds of their observations to Earth, something that has not been accomplished yet by any country.

NASA is looking into testing satellites that are even smaller than microsatellites—nanosatellites. The word "nano" means "one-billionth." For example, a nano-second is one-billionth of a second, if you can imagine such a small span of time! As their name implies, nanosatellites are tiny compared even to micro-satellites, weighing in at only about 40 pounds (18.1 kg). Once again, smaller might very well be better. A linked group of tiny nanosatellites with mirrors attached to them could do the job of one large space mirror. They would also be easier to replace and could be launched cheaply because they are so light and could be released in clusters. A group of satellites could be linked together to make different weapon formations. Some could be stronger, or could shoot farther than others, depending on the situation.

There are other, even more revolutionary plans to make the inner workings of satellites still smaller. Some

NASA HELPING THE NAVY

There is a blurring line between commercial uses of satellite imagery and military uses. NASA, a civilian space agency, is being used to help the military in its war on terrorism. For example, sea-viewing wide field of view sensor (SEAWIFS) and moderate resolution imaging spectroradiometer (MODIS) spacecraft are analyzing weather conditions such as wind, fog, dust, and cloud cover. Military planners use this information when making attack strategies, deciding where to deploy ground troops, and choosing which weapons to use in a certain area.

MODIS's imaging data is also helping to move helicopters through unfamiliar terrain. The information provided by the satellite helps the U.S. Navy tell the difference between fog and clouds, which is important when the aircraft move through foreign territory. For example, U.S. pilots flying helicopters during Operation Enduring Freedom in Afghanistan used this data to negotiate the country's complex and potentially confusing chain of valleys. Special Forces personnel also benefit from information like this. SEAWIFS and MODIS help these soldiers by providing data on temperature, regional dust, rainfall and snowfall levels, and other weather conditions that affect military operations.

NASA scientists are trying to apply nanotechnology to satellite design. Nanotechnology allows for the production of extremely tiny machines that are only visible with the help of powerful microscopes. Some of these tiny machines can then build even smaller microscopic structures. Planners envision launching tiny satellites that can duplicate themselves like a virus. If attacked, they could be programmed to rebuild or replicate themselves while still in orbit! Smaller objects have the additional advantage of being far harder to locate and target. Hostile nations would have to work long and hard to make more sophisticated sensors and radar systems that would be able to track, not to mention destroy, these barely detectable satellites.

GRAVITY

Some space-based weapons currently being considered will really take advantage of the new high ground that space provides. The principle behind them is no different from that of a medieval knight pouring boiling oil upon invaders scaling a castle's walls, though the technology involved is a little more sophisticated. Kinetic energy weapons (KEW) are projectiles that will be dropped from orbiting satellites onto unsuspecting enemies standing hundreds of miles below. Gravity is the enemy of rockets and missiles, or anything launched

Looking for an easier and less expensive way to launch satellites into outer space, scientists are examining the relatively simple design of the electromagnetic rail gun, which fires projectiles using only the power generated by electromagnetism.

from the ground. For KEWs, it is an ally.

A KEW is nothing more than a dense object—a pellet or a rod—that is propelled from space to Earth with the help of gravity and an initial boost from a rocket or a gun. These rods must reach high speeds before they enter the atmosphere; otherwise they will not maintain enough velocity to reach the ground, and they may burn up in the atmosphere. High speed can be generated with the help of booster rockets or with what scientists call a rail gun. This is a long, empty tube lined with rails that conduct electricity. Electromagnets surround the energy rod, which is placed in the empty tube. Massive electrical currents then create a powerful electromagnetic field that launches the rod toward Earth at an extremely high speed.

A KEW could plunge half a mile into a hidden enemy bunker, something even the strongest bunker-busting bomb dropped from a warplane could never hope to achieve. A KEW's speed and destructiveness could be further enhanced with the use of lasers. Aiming a laser along the path of a rod could make it drop at supersonic speeds (faster than the speed of sound), since the laser would burn off the air and heat resistance that would otherwise slow down the rod. This is what U.S. Air Force officials call an air spike.

And so the quest for technological superiority and the race to conquer space have brought us full circle. Like our earliest ancestors, we seem to be clamoring up to the highest ground available to us in order to hurl deadly projectiles down upon our enemies. Yet since the high ground of space can be occupied by any number of combatants, a deadly shootout among the stars may result from launching armed satellites. If that is the case, technological advancement and superiority will not have carried us very far from our primitive origins.

CONCLUSION

Given all the new interest in space-based defenses, there is no telling what strange new weapons will circle over our heads within the next few decades. Technologically, we have come so far in so little time. At the dawn of the twentieth century, in 1903, the first airplane was invented. Its first flight lasted for twelve seconds and carried its inventors, Orville and Wilbur Wright, on a 40-yard (37-meter) journey. By century's end, NASA scientists were considering sending astronauts to Mars and unmanned satellites beyond the solar system. Today, machines orbiting hundreds of miles above Earth's surface let us communicate with other people anywhere in the world, predict hurricanes, spy on our enemies, and even destroy their forces by helping to guide powerful ground-based weaponry accurately toward their targets. What is the next step? Will space become the domain of peaceful communication and research, or the front line of a new kind of warfare, bristling with destructive weaponry?

Up until now, we have used passive military satellites. Their job has been to watch over enemies, take photos,

Powered by a small twelve-horsepower engine, Wilbur and Orville Wright's airplane was the first man-made, heavier-than-air craft to achieve controlled flight. The Wright 1903 Flyer weighed 605 pounds (274.4 kg) and had a wingspan of 40 feet 4 inches (12.3 m). On December 17, 1903, it flew for twelve seconds at 120 feet (36.6 m) in the air.

record information, and ensure that the United States is not surprised by any sudden troop movements, weapons buildup, or missile launch. Now, however, the military is preparing to use satellites in a far more active way, seeking and attacking targets that pose a threat to our national security and that of our allies.

The attacks of September 11, 2001, reminded us that, even with the end of the Cold War and its nuclear menace, it is still a very dangerous world. Supporters of space-based lasers and other weapons say that we need to guarantee our safety on the ground and that of our satellite networks in the sky. Unless we defend ourselves and our satellites from surprise enemy attacks, it is argued, we risk not only losing the balance of power in space but also inviting crippling attacks upon our national security and way of life. Even poorer countries or a handful of well-funded terrorists can do a lot of damage to richer nations by sabotaging a few choice satellites, possibly destroying large segments of our communications, intelligence, and defense networks.

Not everyone thinks that we should start putting weapons—even if only defensive weapons—into space. Peace activists, several of our allies, and many scientists think that no one should seek to gain exclusive control of space. Instead, it should be a frontier for cooperation and improving the human race and life on Earth through the unique research and communications opportunities space offers.

Those opposed to space-based missile defenses are particularly concerned about the danger of setting off a new arms race. Once the United States puts weapons satellites into orbit, the theory is that countries that have the resources will rush to do the same. To maintain an edge over its potential enemies, the United States will then have to create and launch even more sophisticated defenses. Soon enough, the costly and dangerous cycle

The Soviet Union's Mir space station was launched on February 19, 1986, and stayed in space for the next fifteen years. The Mir, which survived the fall of the Soviet Union, has hosted astronauts from many countries around the world, including the United States. The Mir was brought back to Earth on March 22, 2001, by the Russian Space Program.

of military buildup last seen during the Cold War will again spiral out of control.

Even if we do launch a large system of satellite defenses within the next thirty years, our security will not be guaranteed. The system would have to be totally impenetrable to work effectively. Critics argue that a space-based missile defense system will do nothing to protect us from a newer kind of threat—box-cutter wielding terrorists planning on hijacking an airplane or detonating a suicide bomb on a crowded subway train.

The more advanced our systems are, the more ways clever and resourceful people who are bent on destruction can locate and exploit those systems' weaknesses, often using the crudest methods. No defense is perfect or effective for long. Someone will always figure out a way to penetrate it eventually. All of the time and expense will hardly be worth the effort if even one or two nuclear missiles make it by our SBLs or if enemies find a way to attack and disable the weapons satellites.

Some say seizing the new high ground and placing weapons in space is necessary to the defense of our freedom and safety. Others think it represents the turning of yet another bloody page in world history. Perhaps both sides are right. Regardless of the dangers, however, it seems unlikely that this genie, once released, will be able to be put back in the bottle.

It does seem likely, however, that space will simply become an extension of Earth and come to share in its anxious mix of hope and despair, friendship and hatred, peace and war. The way we live on Earth will be reflected back to us from space.

GLOSSARY

actuators Small mechanisms that move parts of a space mirror into their proper position.

Acquisition, Tracking, and Pointing (ATP) The part of a space-based laser that allows it to find and destroy its target.

antiballistic missiles (ABMs) Missiles used to shoot down other missiles, especially nuclear ones.

antisatellite weapons (ASATs) Anything that can be used to destroy or disable a satellite orbiting Earth.

deformable Refers to mirrors, especially space-based ones, that can change their shape.

geosynchronous An orbit that is 22,300 miles (35,888 km) above Earth. Satellites in this orbit travel directly above the equator at the same speed Earth spins on its axis, and therefore remain above a fixed point on Earth. To observers on the ground, it appears that geosynchronous satellites are not moving at all.

infrared Refers to special sensors that show the heat being generated in the world around us.

intercontinental ballistic missiles (ICBMs) Missiles that can travel thousands of miles and can be used to carry nuclear warheads.

microsatellites Small satellites (weighing about 200 pounds, or 90.7 kg) currently in development that will be cheaper and easier to launch than average-sized satellites.

microwaves Radio waves that generate extreme heat in the objects at which they are directed.

multispectral scanning A kind of satellite photograph made of pictures of different wavelengths that can detect and identify small differences in materials or terrain.

mylar A flexible coating material often used to insulate satellites.

nanosatellites Minisatellites of the future that will weigh as little as 40 pounds (18 kg).

optical Having to do with vision.

particle beam A weapon that shoots particles of matter at near-light speed, destroying targets.

payload The load carried by an aircraft or a spacecraft.

projectile Any object that is thrown or shot at a target.

reconnaissance The military exploration of enemy territory.

space-based lasers [SBLs] Orbiting satellites that fire and direct laser beams at nuclear missiles or other targets.

space debris Any useless machinery and material in space that can damage a satellite or other spacecraft upon collision at high speed.

FOR MORE INFORMATION

Boeing Phantom Works
Boeing Space and Communications
P.O. Box 2515
Seal Beach, CA 90740
(562) 797-2020
Web site: http://www.boeing.com

Department of Defense
Directorate for Public Inquiry and Analysis
Office of the Secretary of Defense
Public Affairs
1400 Defense Pentagon
Room 3A750
Washington, DC 20301-1400
(703) 428-0711
Web site: http://www.defenselink.mil

Federation of American Scientists
1717 K Street NW, Suite 209
Washington, DC 20036
(202) 546-3300
Web site: http://www.fas.org

Jet Propulsion Laboratory
4800 Oak Grove Drive
Pasadena, CA 91109
(818) 354-4321
Web site: http://www.jpl.nasa.gov

Kennedy Space Center
Public Inquiries
KSC, FL 32899
(321) 867-5000
Web site: http://www.ksc.nasa.gov

NASA Headquarters
Information Center
Washington, DC 20546-0001
(202) 358-0000
Web site: http://www.nasa.gov

Union of Concerned Scientists
2 Brattle Square
Cambridge, MA 02238-9105
(617) 547-5552
Web site: http://www.ucsusa.org/index.html

U.S. Strategic Command
Public Affairs Office
901 SAC Boulevard, Suite 1A1
Offutt Air Force Base, NE 68123-5455
(402) 294-4130
Web site: http://www.stratcom.mil

MAGAZINES

Air & Space Power Journal
401 Chennault Circle
Maxwell AFB, AL 36112-6428
Web site: http://www.airpower.maxwell.af.mil/
airchronicles/apje.html

Aviation Week & Space Technology
1200 G Street, Suite 922
Washington, DC 20005
Web site: http://www.aviationnow.com/content/
publication/awst/awst.htm

VIDEOS

Eyes in the Sky. Produced by Patrice Andrew and Paul Gasek. Discovery Channel Pictures, 1995.

WEB SITES

Due to the changing nature of Internet links, the Rosen Publishing Group, Inc., has developed an online list of Web sites related to the subject of this book. This site is updated regularly. Please use this link to access the list:

http://www.rosenlinks.com/ls/wesa/

FOR FURTHER READING

Becklake, Sue. *Space, Stars, Planets, and Spacecraft*. New York: Dorling Kindersley, 1998.

Gaffney, Timothy R. *Secret Spy Satellites: America's Eyes in Space*. Berkeley Heights, NJ: Enslow Publishers, 2000.

Graham, Ian. *Planes, Rockets, and Other Flying Machines*. New York: Franklin Watts, 2000.

Graham, Ian. *Satellites and Communications*. Austin, TX: Raintree Steck-Vaughn, 2000.

Kallen, Stuart A. *The Race to Space*. Edina, MN: ABDO Publishing, 1996.

Mellett, Peter. *Launching a Satellite*. Crystal Lake, IL: Heineman Library, 1999.

Vogt, Gregory. *Rockets* (Exploring Space). Bloomington, MN: Bridgestone Books, 1999.

Walker, Niki. *Satellites and Space Probes*. New York: Crabtree Publishers, 1998.

BIBLIOGRAPHY

Bush, President George W. "President Discusses National Missile Defense." *The White House*. Retrieved September 2002 (http://www.whitehouse.gove/news/releases/2001/12/20011213-4.html).

Dao, James, and Andrew C. Revkin. "A Revolution in Warfare: Machines Are Filling in for Troops." *New York Times*, August 16, 2001.

David, Leonard. "ABM Treaty Withdrawal Likely to Boost Space-Based Laser Work." SPACE.com, January 18, 2001. Retrieved March 2002 (http://www.space.com/news/sbl_011218.html).

David, Leonard. "Space Laser Project Heats Up." SPACE.com, November 28, 2000. Retrieved March 2002 (http://www.space.com/businesstechnology/space_laser_001127.html).

David, Leonard. "Star Warriors Eye Space-Based Laser Experiment." SPACE.com, April 7, 2000. Retrieved March 2002 (http://www.space.com/businesstechnology/technology/laser_tech_000407.html).

David, Leonard. "U.S. – Russia Working on Satellite Missile Watching System." SPACE.com, October 24, 2001. Retrieved March 2002 (http://www.space.com/businesstechnology/technology/us_russia_satellite_011024-1.html).

Dupont, Daniel G. "The Real Star Wars." *Scientific American*, June 1999.

Grossman, Karl, and Judith Long. "Waging War in Space." *The Nation*, December 27, 1999.

Grossman, Karl, and Michio Kaku. *Weapons in Space* (Open Media Pamphlets Series). New York: Seven Stories Press, 2000.

Handberg, Roger. *Seeking New World Vistas: The Militarization of Space*. Westport, CT: Praeger Publishers, 2000.

Hitt, Jack. "Battlefield: Space." *New York Times*, August 5, 2001, section 6, p. 30.

Launius, Roger D., and Howard E. McCurdy. *Imagining Space: Achievements, Predictions, Possibilities 1950–2050*. San Francisco, CA: Chronicle Books, 2001.

Lewis, George N., Theodore Postol, and John Pike. "Why National Missile Defense Won't Work." *Scientific American*, June 1999.

Mitchell, Gordon R. *Strategic Deception: Rhetoric, Science, and Politics in Missile Defense Advocacy.* East Lansing, MI: Michigan State University Press, 2000.

Newman, Richard J. "The New Space Race: The Pentagon Envisions a War in the Heavens, but Can It Defend the Ultimate High Ground?" *U.S. News & World Report*, November 8, 1999, p. 30.

Oberg, James. "The Heavens at War." *New Scientist*, June 2, 2001. pp. 26–29.

O'Hanlon, Michael E., and James. M. Lindsay.
 *Defending America: The Case for Limited National
 Missile Defense*. Washington, DC: Brookings
 Institution Press, 2001.
Tirman, John, ed. *The Fallacy of Star Wars: Why
 Space Weapons Can't Protect Us*. New York: Vintage
 Books, 1984.
Vizard, Frank. "Return to Star Wars." *Popular Science*,
 April 1999, pp. 56–61.

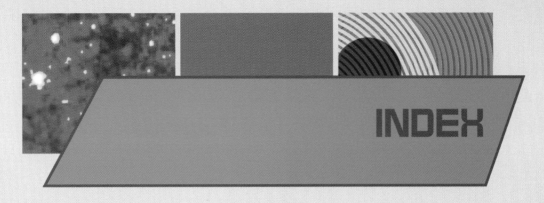

CREDITS

ABOUT THE AUTHOR

Born in Bydgoszcz, Poland, at the height of the Cold War, Philip Wolny emigrated to the United States with his parents at the age of four, settling in Queens, New York. He continues to live and work in New York City as a writer and editor.

PHOTO CREDITS

DESIGNER

Tom Forget